ACE Inhibition in the Management of High Blood Pressure

ACE Inhibition in the Management of High Blood Pressure

Robert J MacFadyen
Consultant Physician/Cardiologist
Raigmore Hospital Cardiac Unit
Old Perth Road, Inverness

MARTIN DUNITZ

Although every effort has been made to ensure that the drug doses and other information are presented accurately in this publication, the ultimate responsibility rests with the prescribing physician. Neither the publishers nor the author can be held responsible for errors or for any other consequences arising from the use of information contained herein.

© Martin Dunitz Ltd 1999

First published in the United Kingdom
in 1999 by
Martin Dunitz Ltd
The Livery House
7– 9 Pratt Street
London NW1 0AE

A CIP record for this book is available
from the British Library.

ISBN 1-85317-730-X

Printed and bound in Italy.

Contents

Introduction

Elevated arterial blood pressure is common and mostly asymptomatic. It is a clinical problem in which treatment, however straightforward, too often fails to deal with the consequences of failure. Despite the huge amount of work around the world drawing attention to this condition and warning of its dangers, in addition to the work on improving its diagnosis, assessment and therapy, vast numbers of patients are left vulnerable to their disease. They are vulnerable to the so-called cardiovascular end points. This seems a remarkably off-hand way of describing sudden death, stroke, myocardial infarction, heart failure, peripheral vascular disease, end-stage renal failure. Yet practising physicians see new events every day in hospitals throughout the world and watch the consequences among families in their respective communities.

The problem, so simple yet meriting continuous restatement, is based in the fact that patients with hypertension rarely if ever feel ill before a cardiovascular event, yet it is at this stage that the opportunities to intervene are so often lost or inadequately grasped. Successful management of hypertension after a non-fatal stroke is in no way a success. Equally, it is understandable that the average hypertensive patient rarely takes a great interest in their own disease, and physicians can easily get lost in the minutiae of diagnostic terminology or cellular physiology.

Without individual commitment to a patient it is easy to lose the goals of seeking out hypertension — defining those who need to be treated and choosing the best treatment for an individual. Although it is on many occasions hard for the physician to persuade the hypertensive patient of the need for detailed assessment, balanced treatment and regular monitoring, equally it is easy to give up in the difficult case where often the patient will default on therapy, appointments and investigations. Although the patient with hypertension can be a 'heart sink' for the physician, it will be only a matter of time before the patient sinks their own heart in the process. Ultimately it is too easy to leave the inadequately treated patient to their own fate with the onset of the many forms of atherosclerotic vascular disease.

This small book is not meant to be yet another all embracing stab at hypertension physiology and pharmacology. Nor does it attempt to demonstrate how much or little the author knows about the literature of hypertension. It is simply a rallying call focusing on the parts of management we all tend to forget and yet which are so important in practice. What it is intended to do is to stimulate our efforts to deal better with this disease. In an age when technology is making huge strides in every walk of life, the population impact of improved care for hypertension in the community would overwhelm the achievements of advanced management of end-stage atherosclerotic disease. At that stage it is too late and huge efforts are required to salvage the patient. We have all the evidence we need to focus on the priorities of this disease, and the simple tools to achieve this right here and right now. The focus is in the community. One of the key tools is the angiotensin-converting enzyme (ACE) inhibitor group of drugs.

The outlook for the patient with high blood pressure

Epidemiology, definitions and who to treat

High blood pressure (HBP) shares with many conditions a fundamental lack of information about the numbers of individuals affected. In the final months of the twentieth century this seems an incredible state of affairs yet is absolutely true. Although *some* medical practitioners *may* know how many patients they have in their practice, integration of these figures across a whole population, even a very small one (such as a small town), is always incomplete. Even population studies give figures that can hardly by generalized. Hypertension is a global clinical problem. Both susceptibility and patterns of hypertension vary around the world yet suitable national and international figures are not available.

Having said all that, and perhaps in reflection of the significance of the problem, no other medical condition could be said to have had the minutiae of measurement so carefully worked out.

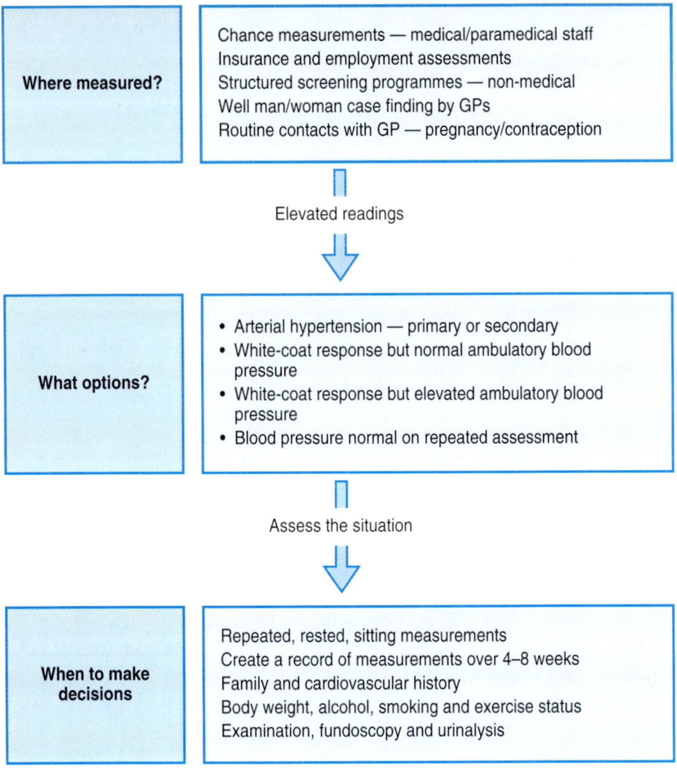

| Where measured? | Chance measurements — medical/paramedical staff
Insurance and employment assessments
Structured screening programmes — non-medical
Well man/woman case finding by GPs
Routine contacts with GP — pregnancy/contraception |

Elevated readings

| What options? | • Arterial hypertension — primary or secondary
• White-coat response but normal ambulatory blood pressure
• White-coat response but elevated ambulatory blood pressure
• Blood pressure normal on repeated assessment |

Assess the situation

| When to make decisions | Repeated, rested, sitting measurements
Create a record of measurements over 4–8 weeks
Family and cardiovascular history
Body weight, alcohol, smoking and exercise status
Examination, fundoscopy and urinalysis |

Figure 1
Where, what and when to measure blood pressure.

The prevalence of arterial hypertension varies in accordance with the technique of assessment:

Practicalities of blood pressure measurement

Tools. Auscultated with a working serviced aneroid (common in family practice) or barometric (common in hospitals) sphygmomanometer; in conjunction with this is a practitioner working to the same criteria of sounds; a serviced automatic device (for home or ambulatory use). Random zero devices which received much attention in years gone by are now for the most part defunct

Who measures blood pressure? Friend; colleague; nurse; general family physician; hospital doctor. It doesn't matter which except to say that all are different and open to measurement technique errors.

How? Seated is preferred, since standing has a marginally higher systolic, and lower diastolic. Standing moving to lying (not sitting) may be useful if a postural change related to nervous system dysfunction may be useful.

Relaxed. Home; surgery; consulting room; casual measurement (shopping mall; pharmacy, etc). Although practicalities dominate our lives and speed and efficiency control most medical activities, blood pressure is preferably not measured after a cervical smear or at the end of a difficult general consultation, and not as the sole reason for contact with the patient. Context is important.

Consistent. This is important: over several readings in the same setting. Hypertension is a chronic disease which almost never requires abrupt decision making.

Classification. Blood pressure is either a primary or secondary problem; much attention to secondary causes despite their rarity still should focus on blood pressure control, eg, the patient has unsuspected Conn's syndrome yet blood pressure control is not markedly disadvantaged.

Age. Blood pressure rises throughout life, as do cardiovascular event rates; thus a given blood pressure measurement is age dependent in terms of cardiovascular risk.

Racial group sampled. Subpopulations vary in the incidence and consequences of hypertension according to their susceptibility and the distribution of other cardiovascular risk factors, eg, African Americans; Japanese Americans; British Asians.

Given these factors, a key group are those who are *not known* to any medical contact as having hypertension. Since hypertension is traditionally thought of as an asymptomatic condition this proportion is no surprise. Are we finding more cases in the community, or less? How many people fall into this category and whether or not it has altered in recent times is unknown. The classical rule of halves is generally held up to account for this phenomenon. Owing to the lack of detailed epidemiology, no one really knows whether this is at all accurate and for what countries it is an overestimate or underestimate. For the true epidemiology of hypertension we rely largely on informed guessing based on statistical samples.

What to do about the white-coat response in hypertension

The white-coat response is an established aspect of measuring blood pressure. On many occasions without obvious impact on the patient or with a simply detectable stress response (eg, tremor, tachycardia, etc) during measurement, transient (up to 2 hours after consultation) elevations of blood pressure are seen in normotensive individuals (white-coat response) or patients with hypertension (white-coat hypertension). The definition of this response has come about with the use of small and unobtrusive ambulatory blood pressure recorders since the early 1980s. All the large epidemiological studies of hypertension therefore contain ill-defined numbers of patients with the white-coat response and normal blood pressure.

How can we put this into the clinical practice equation? Though the long-term significance of a white-coat response in a subject with normal ambulatory blood pressure is as yet unclear, it is associated with minimal, if any, increase in cardiovascular risk. It is probably true that a proportion of patients (~25–30%) with a white-coat response will go on to develop sustained hypertension in later life (> 10 years). Due to the longitudinal ambu-

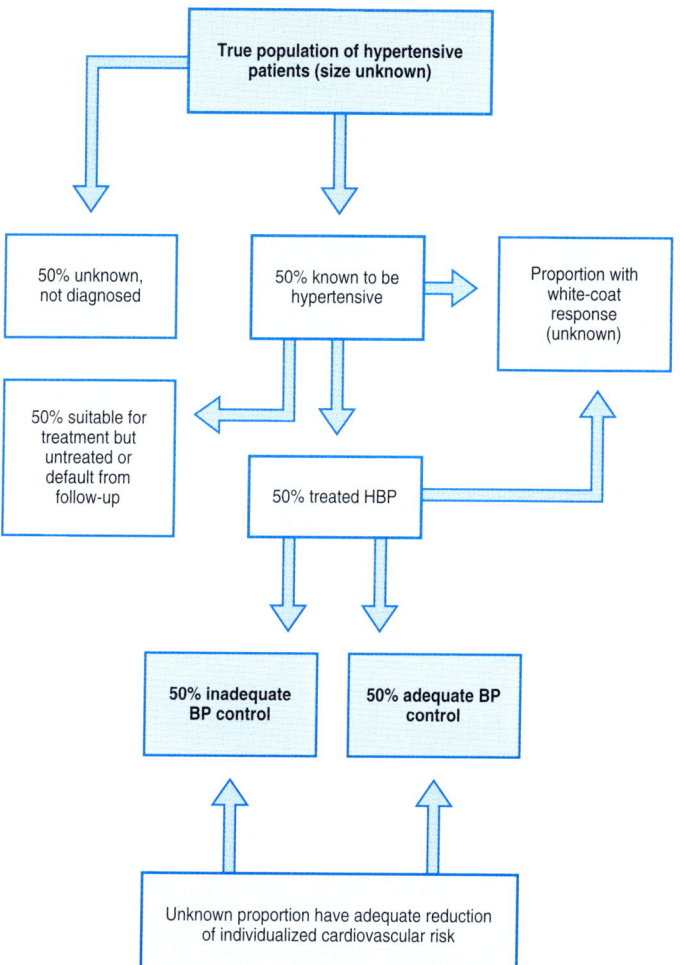

Figure 2
The rule of halves still applies (but half right or half wrong, nobody can say).

latory blood pressure follow-up required to define a white-coat response from ambulatory hypertension the exact proportion off patients affected in this way is still guesswork. Thus a patient shown to have a white-coat response on ambulatory monitoring should probably be regarded as normotensive in the calculation of current cardiovascular risk (see below).

If ambulatory blood pressure monitoring is not widely available, as is most often the case, then repeated clinic or practice blood pressure measurements should be used in calculating initial risk decision making and follow-up (see Figure 3). This results in overestimation of individualized cardiovascular risk and thus a degree of overtreatment. While so many patients remain undertreated or undiagnosed this is hardly a cause for concern.

Relative cardiovascular risk

Though we face many problems in assessing the numbers and definition of high blood pressure, the outcomes are the same. We need to get away from the failures of the past. To agree whether quantitative population epidemiology is accurate or whether blood pressure is a continuous or discontinuous variable is hardly of major importance. Fundamentally we deal with hypertension by focusing on reducing the linked cardiovascular risk. We accept the population epidemiology that the presence of elevated blood pressure (intentionally losing the label 'hypertension' to avoid the number debate) multiplies individual risk.

Thus the epidemiology of hypertension is more about the epidemiology of hypertension among the range of linked cardiovascular events. We take an interest in hypertension not because of its immediate significance, since it is a surrogate measurement only, but because it leads on to mortality and morbidity. We treat hypertension seriously because of the effects of:

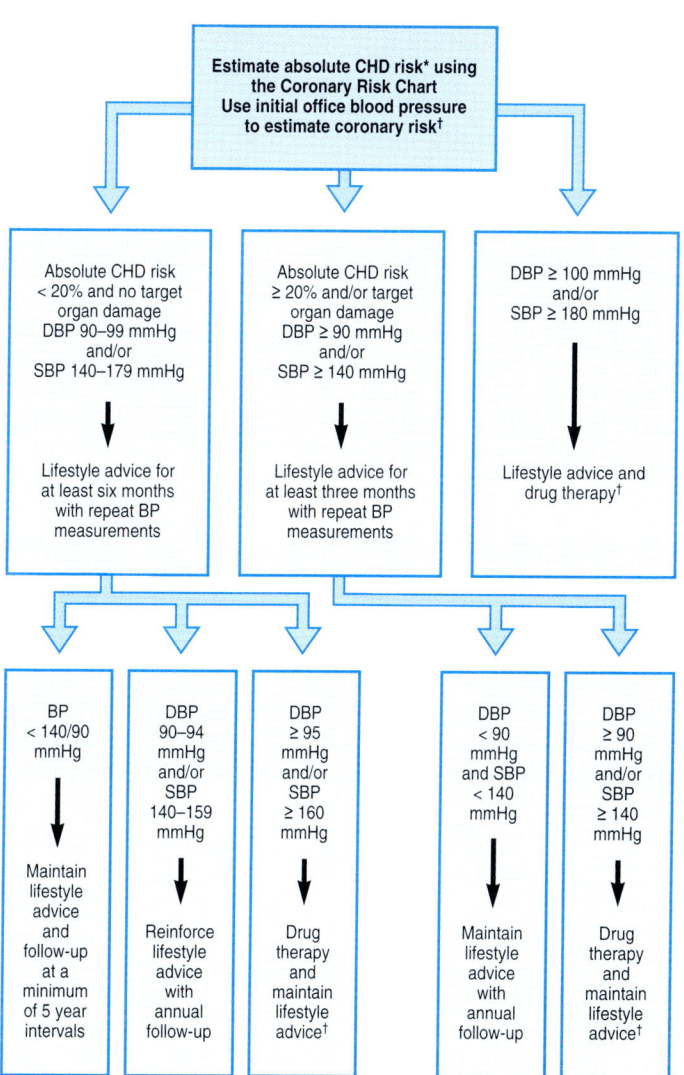

* High CHD risk ≥ 20% over 10 years or will exceed 20% if projected to age 60 years
† Consider causes of secondary hypertension. If appropriate, refer to specialist

Figure 3
Primary prevention guide to blood pressure management. (Reproduced with kind permission from Eur Heart J *1998;* **19***: 1434–503.)*

Many attempts have been made to integrate the assessment of blood pressure with overall cardiovascular risk. We are now at last making significant progress and producing simple ways to guide decision making. Accepting hypertension as only one of several multiplicative risk factors is the first step. We all know the clinical parameters that need to be factored in:

Classic cardiovascular risk factors

- Gender
- Age
- Smoking status
- Body weight (as body mass index weight in kg/height in m^2)
- Glycaemic status (diabetic or not)
- Measured (auscultatory) rested sitting systolic blood pressure
- Lipid status (as Low Density Lipoprotein (LDL) cholesterol to total cholesterol ratio)

These are incorporated in the New Zealand guidelines tables to approximate individualized cardiovascular event risk. These can be used very simply to address the need to intervene in any given individual and the impact of intervention, ie, the need to try harder which is so often a problem in the hypertensive

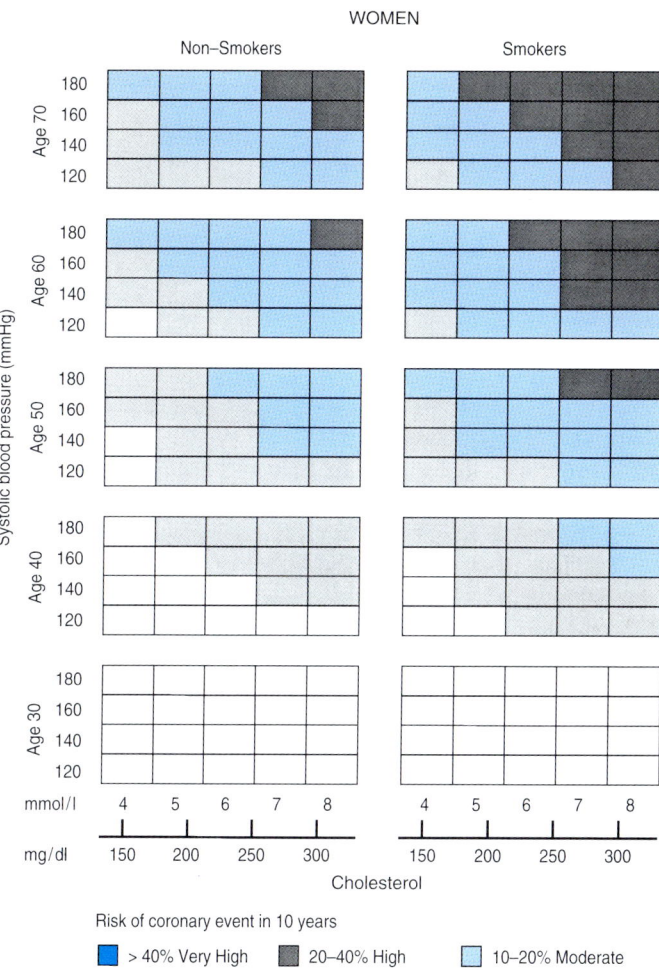

Figure 4
The New Zealand guidelines for assessment of individualized cardiovascular risk.
(Continues overleaf).

11

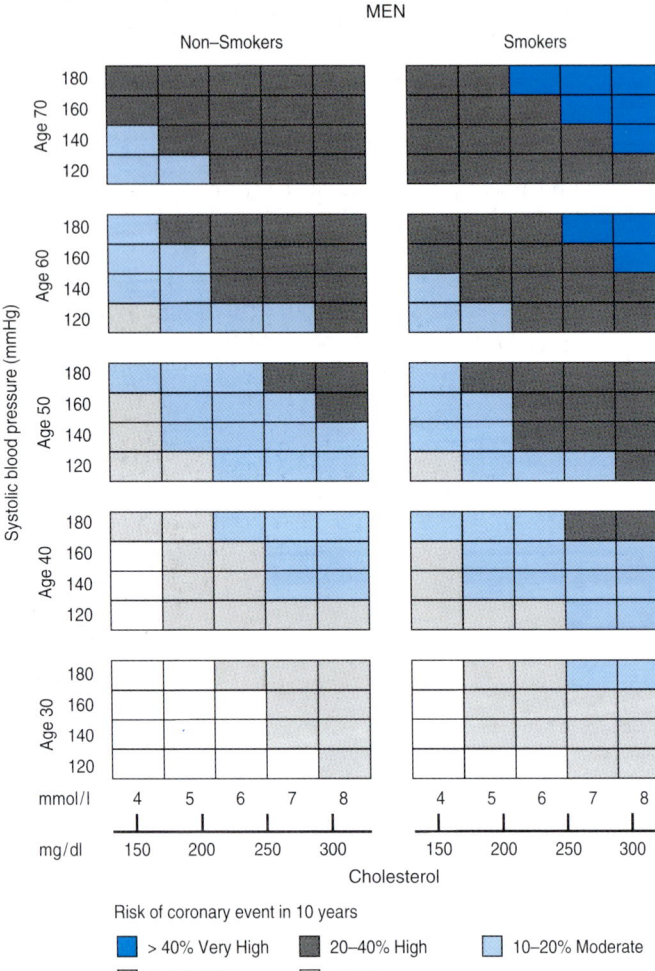

MEN

Risk of coronary event in 10 years

- ■ > 40% Very High
- ■ 20–40% High
- ■ 10–20% Moderate
- ■ 5–10% Mild
- □ < 5% Low

Figure 4
*The New Zealand guidelines for assessment of individualized cardiovascular risk.
(Continued). (Reproduced with kind permission from Eur Heart J 1998: **19**:
1434–503.)*

How to use the Coronary Risk Chart for Primary Prevention

The chart is for estimating coronary heart disease (CHD) risk for individuals who have not developed symptomatic CHD or other atherosclerotic disease. Patients with CHD are already at high risk and require intensive lifestyle intervention and, as necessary, drug therapies to achieve risk factor goals.• To estimate a person's absolute 10 year risk of a CHD event find the table for their gender, smoking status and age. Within the table, find the cell nearest to their systolic blood pressure (mmHg) and total cholesterol (mmol/l or mg/dl).

- The effect of lifetime exposure to risk factors can be seen by following the table upwards. This can be used when advising younger people.

- High risk individuals are defined as those whose 10 year CHD risk exceeds 20% or will exceed 20% if projected to age 60.

- CHD risk is higher than indicated in the chart for those with:

  — Familial hyperlipidaemia

  — Diabetes (risk is approximately doubled in men and more than doubled in women)

  — Those with a family history of premature cardiovascular disease

  — Those with low HDL cholesterol. These tables assume HDL cholesterol to be 1.0 mmol/l (39 mg/dl) in men and 1.1 (43) in women

  — Those with raised triglyceride levels > 2.0 mmol/l (> 180 mg/dl)

  — As the person approaches the next age category.

- To find a person's relative risk compare their risk category with that for other people of the same age. The absolute risk shown here may not apply to all populations, especially those with a low CHD incidence. Relative risk is likely to apply to most populations.

- The effect of changing cholesterol, smoking status or blood pressure can be read from the chart.

patient. Risk can be reassessed when one or more (or none) of the modifiable factors has been changed (none of us get any younger; few of us change sex no matter how hard we try). This is applicable no matter which or how the parameter changes, eg, a 10 kg loss in body weight; a 3 mmol/l reduction in total cholesterol with weight loss; a 5 mmHg fall in systolic pressure with a low salt diet etc or indeed the impact of drug therapy.

The threshold of intervention is open to debate. For asymptomatic individuals before an event (ie, the primary prevention group) it is currently felt that an event rate of 20% over 10 years requires modification by one or more interventions to reduce overall risk as far below this level as is possible. To reduce it below this is not cost effective. Equally, patients who have already sustained an atherosclerotic vascular event, though not included in such tabulations, can generally be regarded as at high risk of recurrence (> 20% risk of recurrence after 10 years) and thus merit *any* interventions appropriate to reduce risk.

Appropriate risk reduction interventions may include blood pressure reduction but equally might focus on other linked modifiable cardiovascular risk factors, such as weight, glycaemic status, lipid management, smoking habit, etc. One or all modifiable factors might be targeted in any given individual, dependent on their response. For example, a patient may stop smoking cigarettes, gain 10 kg in weight and still reduce overall cardiovascular risk by 5%.

One aspect of this equation which is a greatly undervalued intervention, particularly in the hypertensive population, is smoking cessation.

Remember:

Key points on smoking and health

- Smoking is pleasurable and rewarding for smokers
- Smokers who claim to have stopped often have not stopped completely or even not stopped at all
- Positive advice about the health benefits of stopping smoking (reduces cancer risk; reduces cardiac risk; reduces respiratory incapacity; reduces personal financial costs) is effective, especially when given by a medical practitioner, but *minimally* so
- Smoking is a physical as well as mental dependence (addiction) on nicotine
- The tar and not the nicotine content of cigarettes is primarily responsible for the adverse cardiovascular effects of smoking
- Provide nicotine replacement to break the smoking habit as a positive decision

---

Continued smoking antagonizes any effect on blood pressure control, limits the impact of drug therapy and obviously worsens the risk of linked respiratory disease and cancer rates which ultimately make cardiovascular management pointless. Though information on the health benefits of stopping smoking are the baseline for asymptomatic adults, more aggressive anti-smoking management is required for hypertensive patients, often in the form of nicotine replacement therapy.

The principles and practice of smoking cessation

Principles and steps:

1. Cigarette smoking must be replaced completely but can be substituted gradually

2. Nicotine replacement must be provided in quantities sufficient to suppress cigarette craving and in a form that is most acceptable to the smoker (gum, dummy cigarette, transdermal patches, etc)

3. Smoking should always be reduced with the onset of replacement

4. Replacement therapy may need to continue for some time after the cessation of all cigarette intake while gradually reducing the level of replacement

5. Although chronic therapy is not envisaged, some smokers may require years of therapy and some may never be able to stop

In summary, blood pressure management is shifting from a previous 'culture of (cut-off) numbers' to a considered judgement in the context of cardiovascular risk for both the decision to treat and where to consider further treatment. This directs us towards the right patients for whom to consider therapy that underlines all aspects of successful management.

Goals in the management of hypertension

Preventive cardiovascular care: how successful?

The most important aspect of management of hypertension is the reduction of future cardiovascular events. In recent years cardiovascular risk reduction strategies have been targeted at those individuals at high risk. Hypertension is a major player in this assessment. While the exact point of initiation of treatment strategies (10–20% 10 year risk of an event) is still the matter of debate dependent on the screening system used, the overall strategy is one of reducing *all* simple risk factors in any individual patient (see page 8).

With this in mind until recently the reduction in coronary vascular disease and myocardial infarction has been felt to be less than expected with blood pressure control when examined in the large longitudinal follow up studies of treated hypertensive patients. The reasons for this are very complicated but may relate to an under-appreciation of linked cardiovascular risk factor management (for example lipid control, glycaemic status, concomitant and continued smoking despite BP control) or even the effects of unrecognized white-coat response. Most of the large population studies in hypertension were completed before the early 1980s when ambulatory blood pressure assessments allowed the nature of the white-coat response to be more fully appreciated. While this group may have some minimal increase in vascular risk and even a degree of progression to fixed arterial hypertension in later life, they will tend to under-estimate the effects of treatment as they dilute the event rate in unselected hypertensive patients. They will appear as uncontrolled or controlled (in some cases) hypertensive patients who have fewer events than equivalent true hypertensive patients (regardless of treatment effects).

Hence the strategy evolves where an overall approach is taken and in some instances rather than in all cases this may focus on one issue such as blood pressure control. For example a grossly overweight, hypertensive, non insulin dependent diabetic who achieves significant weight loss (eg, 25 kg), stops smoking and shows resolution of diabetes and secondary dyslipidaemia yet maintains a blood pressure of 155/95 might no longer require pharmacological treatment of hypertension due to the significant reduction in overall risk. Equally, a 60 year old slim, normoglycaemic, normo lipidaemic, non smoking woman with little blood pressure reduction from 185/100 on beta blocker monotherapy will require further assessment and treatment to resolve her vascular risk. Blood pressure control would be the important target in this case.

While these strategies are only emerging clearly in the early to mid 1990s we have few true markers of the impact of these treatments. Necessarily the view must be taken on a very large scale and encompass many varied populations with differing genetic risk of cardiac events and hypertension. Few organizations or governments have committed themselves to supporting population surveys. One such effort has been initiated across the world by the World Health Organization. The world wide MONICA survey (simply a monitoring exercise defining the patterns of care, cardiac risk factors and outcomes) provides regular vital data reports. Their most recent observations probably reflecting patterns of care up to the mid 1980s, given a lag time to progress to events, do not support the role of risk reduction strategies on the global prevalence of cardiovascular disease events. This may simply reflect the lag time of the new strategies and given the scale of the project should not prevent or inhibit comprehensive risk factor management.

Protecting the heart, kidneys and blood vessels in HBP

End organ damage is a critical event in the prognosis of the hypertensive patient. Recognition of this can be made at a variety of levels but the end organ primarily affected is the vascular endothelium. Though the concept of a single cell layer of the blood vessel as an organ system is relatively new, it is one level where practising physicians can share a common purpose with cell or molecular scientists. Research markers of endothelial and vascular dysfunction in hypertensive patients are numerous and include elements of:

Vascular damage in hypertension

- Impaired smooth muscle contractile and relaxation response
- Amplified response to endogenous physiological pressor substances
- Impaired physical compliance
- Altered resistance capacity
- Increased thrombotic tendency
- Amplified inflammatory response
- Altered response to mechanical stress

Some or all of these factors are evident not only in patients with hypertension but also in some states regarded as pre-hypertensive, such as the as yet normotensive offspring of hypertensive parents. Yet these are not used clinically. Evidence of damage to the end organs of heart, blood vessels, kidneys is an important marker for practical management. This can be recorded easily at the time of the initial assessment, ie:

The initial hypertensive patient assessment and end organ damage

From the history

1. Active atherosclerotic disease: ischaemic heart disease (IHD); peripheral vascular disease (PVD); previous stroke syndrome

From the physical examination

1. Cardiomegaly

2. Any arterial bruits: neck; aortic; renal; legs; arms

3. Advanced retinopathy at fundoscopic examination (for example haemorrhage; retinal ischaemia or exudates)

From basic lab assessments

1. Electrocardiogram: for signs of ischaemia or LVH; conduction anomaly etc.

2. Urinalysis: for proteinuria and more rarely haematuria

3. Serum and/or urinary biochemistry: signs of impairment of renal function and/or hyperuricaemia

4. Chest radiograph: cardiomegaly

The place for echocardiography in hypertension assessments

1. Where there is concomitant heart disease (eg, murmur or previous infarct history) to define echo LVH; wall motion and thickness; valve function etc.

2. Where the ECG or CXR is *abnormal*, to define cardiac morphology

---

The presence of end organ damage suggests a later stage in the process of hypertensive damage and therefore not only removes all diagnostic doubt but also casts greater emphasis on the need for effective treatment owing to the associated poor prognosis that evidence of end organ damage entails. The patient should be informed of these findings and that successful blood pressure control is effective in reversing the impact of this damage. This should be used as a positive educational tool to reinforce the effectiveness of the consultation. Identification is this simple and gives important information for the patients and the physician to consider together in making treatment choices.

Symptoms in hypertension before vascular events: a misunderstood area

There is a paradox in hypertension linking symptoms to this ostensibly asymptomatic condition. In the past and present many practitioners have linked simple (usually ill-defined) headache or non-specific visual anomalies (even refraction errors!) to the presence of hypertension. Formal studies do show increased symptomatic complaints and higher incidence of depression, headache, nocturia, tinnitus, dizziness, epistaxis and impotence. Much of this is felt to be self-selection bias in patients seeking medical care whose blood pressure elevation is detected by chance. Although the evidence for causal links between elevated arterial blood and non-specific headache or visual refraction errors is minimal, some hypertensive patients can experience symptoms as a result of their disease.

Hypertension is predominantly an asymptomatic state yet symptoms are well recognized albeit very rarely and in extreme circumstances. For example in the pregnant woman paradoxically unprotected by previous hypertensive vascular damage, relatively rapid and in some instances small elevations in blood pressure lead to encephalopathic damage and secondary seizures. More rarely in the accelerated or so-called malignant phase of essential hypertension a similar pattern occurs. This is now almost never seen in westernized cultures, owing largely to the prevalence of general medical care, blood pressure measurement and at least partial treatment before the onset of an ischaemic brain syndrome.

Evidence also links the presence of hypertensive left ventricular hypertrophy with the impaired exercise capacity and possibly outcome in hypertensive patients. Owing to the vascular changes in hypertension, the normal stroke volume and heart rate increase with exercise are reduced. There is a consistent blood pressure overshoot on exercise in all hypertensive patients. Exercise capacity can be up to 30% lower than in age-matched controls. It is not yet clear if these changes relate to alterations in body mass and insulin resistance. The blood pressure response to exercise may be an independent marker of cardiac events.

Lifestyle and pharmacological treatment

The critical issue for the so-called 'lifestyle' or non-pharmacological interventions is their quantitative impact and associated with this the evidence for efficacy. Though blood pressure is now considered only one aspect of individualized cardiovascular risk, as detailed above, it is nonetheless an easy and relatively reproducible surrogate measure. The lifestyle modifications are simple and now known to a reasonable proportion of the general population:

Non-pharmacological treatment

- Alcohol intake: reduce below the accepted norm: males 28 Units per week; females 21 Units per week
- Reduced salt intake: no added salt; in selected motivated intervals a low sodium diet may be very effective (eg, 40 mmol/l sodium as chloride; 20 mmol/l potassium)
- Exercise level: increase general level to include at least one daily episode of exercise of at least 10 min duration which is sufficient to induce breathlessness
- Body weight and caloric intake: reduce to achieve a body mass index < 25 kg/m^2
- Stress management: goals here must be self-selected but can include stress management or alternatively assertiveness training; development of coping skills; meditation techniques including the use of physical strategies (eg, yoga) and/or relaxation and biofeedback devices. Many of these issues are workplace related and apply more to males than females
- Smoking cessation: this is probably mandatory for all cardiovascular patients and its impact dominates most of the other interventions (see above)

All the above factors are subject to variable documentation by individual patients. Understandably, few asymptomatic individuals will admit to 'excessive' smoking or alcohol intake. Equally, exercise levels in Western societies among middle-aged and unselected subjects are minimal to zero with few exceptions. Despite the positive effects of weight reduction on self-esteem and functional capacity from even modest exercise (walking rather than transport for short distances), population uptake of simple dietary advice is uniformly poor. Reduction in body weight should be obvious to all as a fundamentally good intervention (at many levels) but is one of the least attainable at an individual level. Sodium content in westernized diets are chronically elevated by virtue of additions made during food processing. Practical modification of salt intake is usually fairly straightforward and requires avoidance of processed foods and cessation of added salt as seasoning for prepared food or in cooking food.

The overall impact of any one of these lifestyle interventions on blood pressure is small in the individual patient (1–5 mmHg) and compliance is difficult to assess. None of these interventions is known conclusively to reduce cardiovascular risk. However, the gain in terms of motivation and self-control is well worth the effort. Moreover, simple statistics support clinical observation that an appropriately motivated individual patient can show dramatic changes, eg, in losing weight, reducing alcohol or removing added salt and thereby lowering blood pressure substantially.

Whether these are exceptions (literal outliers on a graph of hypertensive responders) rather than the rule is immaterial. Since the reduction of cardiovascular risk and blood pressure management are almost always long-term goals, it is more than appropriate to persevere with non-pharmacological techniques as an initial step provided it is safe to do so. Safe application of non-pharmacological techniques generally requires an absence of end organ damage.

Optimized drug treatment for individual patients

Individual hypertensive patients require long-term therapy and for most this will be lifelong. Though antihypertensive drug withdrawal has been documented in small numbers of individuals, this probably relates to inaccurate initial diagnosis. Patients must understand the need for therapy and its goals. They should be informed about the possible side-effects of individual treatments and helped to understand the benefits of different classes of treatment. A portion of the variability in blood pressure response and tolerability may even be genetic in origin. Although the definition of responsible loci is underway, particularly with respect to hypertension, studies are at a very early stage and as yet have no practical application.

The main problem for many patients with hypertension is still inadequate treatment. In the US National Health and Nutrition

Examination Study survey, patients with treated hypertension had blood pressure values only slightly lower than untreated hypertensives and remained higher than the normal population. A package of treatments must be determined on an individual basis that the patient agrees to, understands, complies with and ideally can monitor along with their physician. The keys are:

- The best assessment

- The best drug

- The correct dose

Monotherapy and combination treatments

Clinic blood pressure is the target. Treatments should normalize observed readings in that consistent setting. Drug treatment should be considered in addition to the non-pharmacological approaches above. Effectiveness is judged on the basis of cardiovascular risk reduction. There is generally no blood pressure that will be too low so long as the patient remains asymptomatic, and much evidence suggests the lower the blood pressure the better.

Individual drug therapy should be selected and given individual patient trials for an adequate time to allow the response to emerge (at least 1 month). Most treatments become more effective over repetitive dosing, but the initial response can be used to predict overall effect. Repetitive trials of monotherapy are preferred but in practice these result in control rates of only 40–50%. This relatively poor result owes to inadequate information about patients and poor drug selection with regard to tolerability as much as blood pressure lowering efficacy. Dose titration of monotherapy is little used, probably again because of the association with an increased incidence of dose-related side-effects (see Figure 5).

Individualized therapy requires trust and communication

- Understanding
  - — Diagnosis
  - — Risk and outcomes
  - — Timescale
- Treatment and monitoring options
  - — Lifestyle
  - — Pharmacological
  - — Need for repeated measurements
  - — Benefits of treatment may be at cost of some side-effects

Figure 5
Individualize the antihypertensive treatment.

An alternative strategy has been proposed in order to achieve adequate blood pressure reduction, in which low doses of two drugs are employed to minimize the risk of adverse events that could promote non-compliance. Drug selections for these combinations should be chosen to provide a synergistic response (that is, greater than additive whenever possible).

The role of compliance and non-compliance in outcomes

It is obvious that the compliance with any treatment regimen defines its likely outcome. Poor compliance leads to poor outcome. Hypertension is the classical example of a chronic asymptomatic condition where compliance should dominate practical management. Yet published studies are remarkably rare and this is often a grossly neglected area of the clinical consultation despite most practitioners' awareness of its importance.

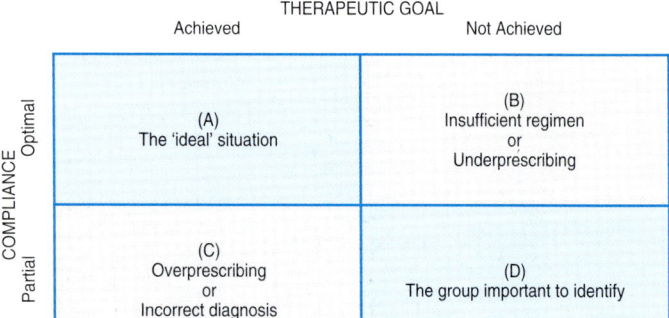

THERAPEUTIC GOAL

	Achieved	Not Achieved
Optimal	(A) The 'ideal' situation	(B) Insufficient regimen or Underprescribing
Partial	(C) Overprescribing or Incorrect diagnosis	(D) The group important to identify

COMPLIANCE

Figure 6
Possible outcomes in therapeutic management, dependent on compliance.
(Reproduced with permission of S. Karger AG, Basel from Rudd et al. Cardiology
1992; **80(1)***: 2–10.)*

It is estimated that hypertensive patients are only 50% compliant with their prescribed regimens by pill count. If this occurs within the therapeutic goal in terms of blood pressure there is less of a problem. Problems occur where failure to lower blood

Class of drug	Total exposed	No of users	1st month
Diuretics	12 157	5 171	4 205 (81)
Beta blockers	9 348	3 615	3 075 (85)
Calcium channel blockers	7 176	3 244	2 539 (78)
ACE inhibitors	5 811	2 710	2 232 (82)

Table 1
Monthly continuation rates for new courses of treatment with antihypertensive
drugs. Figures are numbers (percentages in brackets) of patients continuing
treatment by month (including those continuing with drug identified as new
course of treatment). (This table was first published in Jones et al. BMJ *1995;*
311: *293–5 and is reproduced with permission of the BMJ.)*

pressure is due to unacknowledged partial compliance or total non-compliance (although the latter is relatively rare), since this often leads to prescribing alterations (generally increases in drug dosage or additional drug therapy) and, consequently, worse compliance.

Equally, multiple drug dosing (two, three or four times daily) can often result in timing errors (delays or omissions) in dosage — worsened if more than one drug is involved. This is increasingly important in controlling the antihypertensive response over a 24 hour period. Poor compliance with multiple drug dosing must be recognized as a contributory factor to failure to achieve this goal. In hypertension we start from a bad baseline, since 50% of patients default from follow-up and 50% of those started on treatment drop out in the first year (see Table 1). This poor record seems to affect all types of antihypertensive drug.

However, the basic patient management solutions are simple. The biggest factor in this situation centres firmly on the doctor–patient relationship. Resolution is dependent on atten-

2nd month	3rd month	4th month	5th month	6th month
3 211 (62)	2 668 (52)	2 372 (46)	2 230 (43)	2 131 (41)
2 533 (70)	2 215 (61)	1 983 (55)	1 858 (51)	1 779 (49)
1 877 (58)	1 598 (49)	1 450 (45)	1 367 (42)	1 323 (41)
1 760 (65)	1 501 (55)	1 367 (50)	1 275 (47)	1 211 (45)

tion to patient communication, level of trust, education about the goals of treatment and therapies employed, and ongoing supervision including the assessment of compliance. Appreciation of issues pertaining to the individual patient's quality of life is important. For example, some patients will tolerate a chronic minor side-effect if they appreciate the benefits of therapy, whereas if they remain uninformed they may covertly reject treatment or default completely from follow-up.

Treatment compliance is a neglected issue in hypertension therapy

Non-compliance: the results of non-detection
- Ineffective therapy but this may not occur in some
- Increased circadian cardiovascular risk if selected doses are missed which coincide with maximal risk of cardiovascular events
- Bad, deteriorating or even complete failure of patient–doctor relationship, where both sides may suspect but neither addresses the issues

Non-compliance: routes to avoidance
- Open communication of the issues
- Patient understanding of the needs and goals of therapy
- Patient understanding of the options available and balances required
- Honesty and trust
- Exploration of alternatives
- Monitoring of compliance and efficacy (for example, pulse rate for beta blocker)
- Consider non-compliance or partial compliance as a primary cause of treatment failure and explore the issues on an individual basis

The patient's view and the physician's options

Patients' views on the treatment of their blood pressure problem are not often solicited. Not surprisingly most prefer to feel that they do not in fact have hypertension, which makes confirmation of the diagnosis and explanation of the findings an important starting point. Thereafter it is fair to conclude that issues of quality of life dominate the patient's viewpoint, and resolution of the blood pressure elevation (but now more correctly the cardiovascular risk assessment) dominates the physician's viewpoint.

Lifestyle alterations seem initially attractive to many patients and, as stated above, should always be actively considered. However, they are *not* easy options and most fail through a combination of simple failure of adherence in the long term and inadequate quantitative impact.

On the other hand, although drug therapy is easy to prescribe it must always be part of the 'whole patient' package of cardiovascular risk and done with patient involvement in mind. Most patients when asked will always suggest that they want 'the best' treatment. This is hardly a surprising response. What is the best therapy will be an individual assessment. Quality of life usually means freedom from all adverse effects of treatment yet a balance may be struck based on a simple description of the benefits of therapy tailored for the individual. What factors need to be considered in the physician's selection?

General issues in drug selection

Drug treatment should, with few exceptions, be once daily with a 24 hour profile of effect. This can be achieved by the use of drugs that have naturally protracted duration of effect or that have been prepared in pharmaceutical dosage forms compatible with this priority. Blood pressure control is imperative in patients in whom end organ damage has been identified. However, the maximal response to any given dose can be given time to emerge and a period of dose titration is acceptable. Repetitive assessment should be agreed between patient and physician until set goals are met for blood pressure within the individualized cardiovascular risk assessment.

Options in drug selection

The available key agents are well known and minor drug classes such as clonidine etc, or older agents such as methyldopa, reserpine etc, are not relevant to the majority of cases. Selection of drug class can be guided by patient characteristics. Side-effect profile is generally important, as is the individual patient characteristics. The first-line agents are: thiazide diuretics; calcium channel blocking drugs; beta adrenoceptor antagonists; and the ACE inhibitors. These are used as monotherapy. The symptomatic effects of the older drugs are well defined and understood by almost all practitioners and indeed by many patients. They can prove problematic in some patients but in many, a *trial* of therapy is merited at low dose to assess tolerance, since efficacy may be achieved without encountering side-effects. Equally, many patients can have non-specific side-effects and occasionally more dramatic responses (such as vagal syncope) *unrelated to active drug* (Jarisch–Bezhold reflex).

Thiazide diuretic therapy remains the cheapest option, with a modest blood pressure lowering effect evolving over 1–2 weeks of therapy. Combination preparations or dose increments should be avoided, since electrolyte disturbance is common in elderly patients. Neither loop diuretic drugs nor aldosterone antagonists should be used in essential hypertension although the former may have some role for patients with marked renal impairment and the latter in secondary hypertension related to aldosterone excess. These are very rarely encountered in community practice.

Beta blockers are generally best avoided in proven obstructive lung disease and all smokers (even those trying to stop) although many can tolerate this therapy. Younger patients who are still physically active and those wishing to exercise to lose weight or as part of their non-pharmacological therapy will also more often run into problems with non-specific lethargy due to beta blockade. Impotence can be problematic in some male patients. Despite relative contra-indications the diabetic population tend to tolerate these agents reasonably well and again, despite contra-indication evidence that beta blockade markedly worsens, claudication is at best modest.

One area where traditional caution should remain, however, is the use of beta blockade in patients with known cardiac failure. Although better evidence of the benefits of beta blocker therapy in heart failure is currently emerging, this is still an area where these drugs should be avoided in family medicine for reasons of safety. Safety is one good reason that these agents might be preferred in hypertension during pregnancy.

The vasogenic oedema of the calcium channel blocking drugs is generally more pronounced with short-acting drugs or formulations and to an extent has been attenuated by sustained release dosage forms. Though still affecting only a minority of patients, it can be very problematic when it does occur and still happens with long-acting drugs such as amlodipine. More worrying have been recent suggestions that these agents (both long-acting and short-acting dihydropyridine classes) are associated with an adverse outcome in hypertensive patients with active ischaemic heart disease. Although still being worked out in prospective follow-up, the Nurses Health Study and subgroups of a large trial in diabetic hypertension, both reported in 1998, showed worrying confirmation of a previous meta-analysis in 1996. In the latter study, treatments involving the long-acting calcium channel blocker, nisoldipine seemed to be associated with increased cardiac mortality.

The angiotensin-converting enzyme inhibitors are considered in detail below; their profile of activity and efficacy make them important candidates for most patients but particularly those with diabetes, ischaemic heart disease or hypertensive renal impairment. Care (but not necessarily exclusion — see below) is required in those patients who have signs of widespread atherosclerosis (eg, arterial bruits on examination) and/or renal impairment evident on biochemistry. ACE inhibitors should not be used in pregnancy and are probably best avoided in younger female patients unless pregnancy can be excluded (for example with previous sterilization).

The renin angiotensin aldosterone system in HBP and its response to ACE inhibitor treatment

The renin angiotensin aldosterone system (RAAS) is a fundamental piece of physiology which exists is most species of animal down to the primitive cartilaginous fishes. It exists to support blood pressure with variable fluid volume in the circulation and reacts rapidly to constrict blood vessels and retain sodium from the renal filtrate. The main controlling factors are angiotensin II and aldosterone. It has many other more sophisticated cellular and regional effects on the cardiovascular system (Figure 7).

Although at one time it was thought to be entirely based on the hepatic production of angiotensinogen, the kidney production of renin, and the pulmonary production of ACE which produced angiotensin II, the RAAS is now known to be widespread in the tissues of blood vessels, heart and brain, to name a few of the main systems affected.

One of the fundamental effects of ACE inhibition is that it interacts with the system at multiple levels in the cell and the circulation. By virtue of the importance of the RAAS in the control of blood pressure the effects of blocking the system are dependent on how active the system is in the generation of high blood pressure (Figure 8). There are two instances when the response can be particularly marked:

Where ACE inhibitors have a pronounced effect

- Renovascular stenosis: this is most often the simple form accompanying generalized atherosclerosis or more rarely when the renal vessels are narrowed by muscular dysplasia
- Intravascular fluid depletion: this can be produced by many mechanisms from chronic diuretic therapy, diarrhoeal illness, profound exercise, hyperthermia, etc.

Both of these instances can usually be predicted and the nature of the response can be anticipated, since it owes everything to the withdrawal of angiotensin II. Blood pressure can fall markedly, producing postural symptoms, and renal function can be impaired. This response can be generated in any person, depending on circumstances (eg, fluid and sodium balance), the ACE inhibitor selected and the dose used.

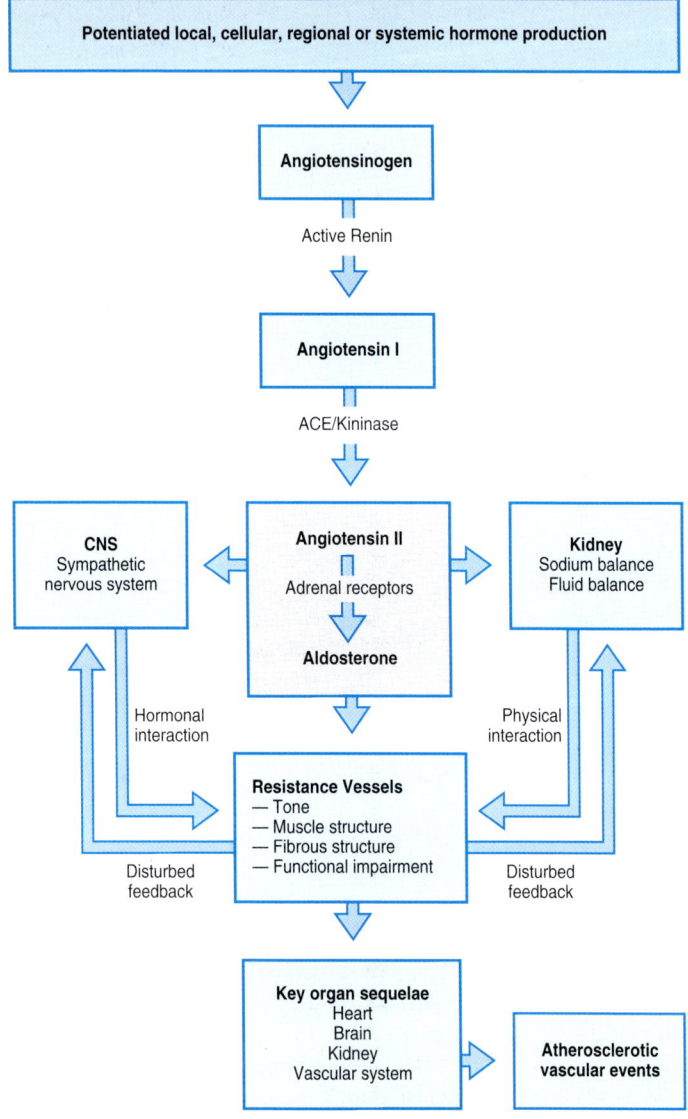

Figure 7
The renin angiotensin aldosterone system in hypertension.

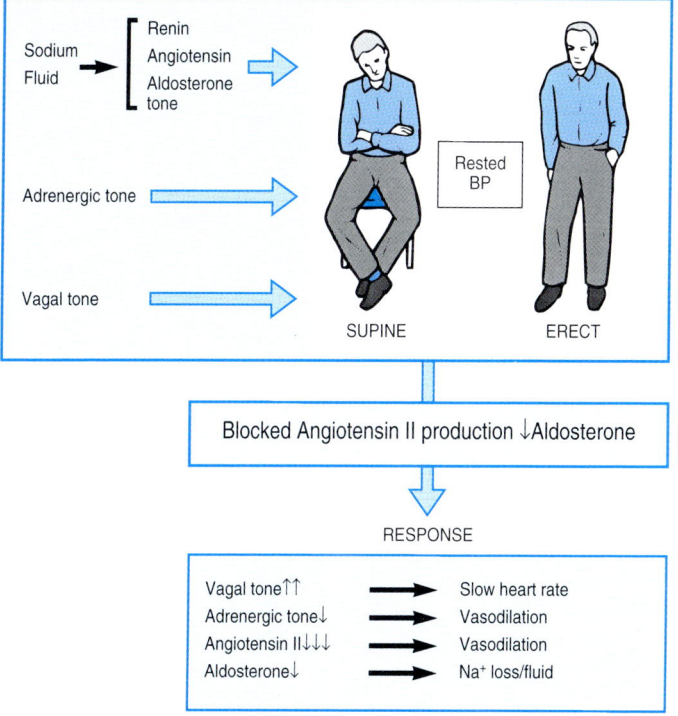

Figure 8
Increased contribution from RAAS implies an increased ACE Inhibitor response.

Such changes do not need to occur, since simple patient assessment and careful selection of the agent and dose used will avoid this pattern almost without fail. Correction of diuretic induced fluid depletion beforehand can be simply effected by discontinuing the drug. Despite the risk of further renal impairment, ACE inhibition is an effective medical treatment for hypertension due to renovascular disease provided that the problem is unilateral. Bilateral disease is a contra-indication to treatment as effective renal failure can be induced.

Cardiac mass and the left ventricle in treated hypertension: remodelling the hypertensive heart

Involvement of the heart in hypertension is a significant finding. In most patients in primary care this will be evident either on history (for example ischaemic chest pain implying coronary disease involvement), or by examination (for example cardiomegaly detected clinically, on chest X-ray, by ECG or echocardiogram). Findings of hypertensive ventricular hypertrophy increase the risk of myocardial infarction by five-fold greater than that of the hypertensive patient without ventricular hypertrophy. This is a powerful predictor of morbidity and mortality in patients with hypertension. Importantly the RAAS is centrally involved in mediating this hypertrophic response (Figure 9).

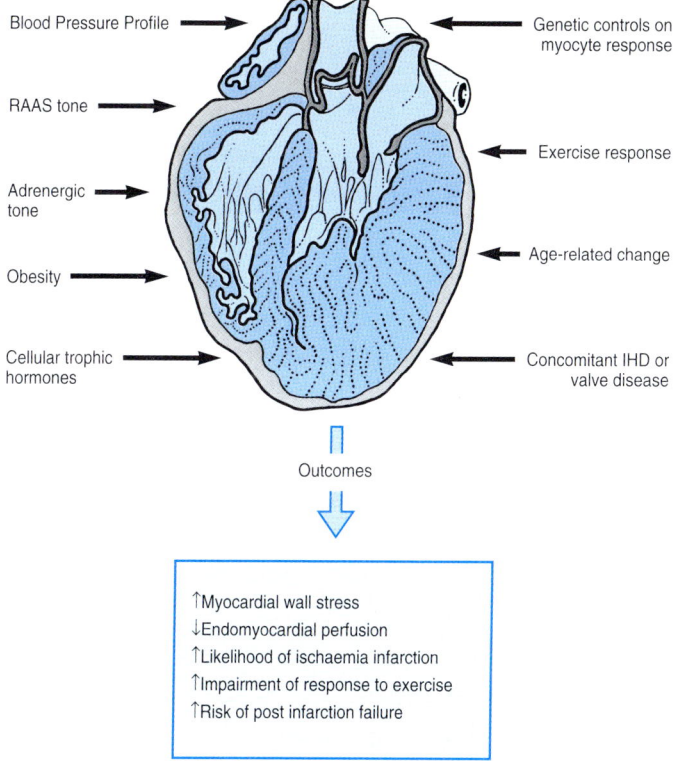

LEFT VENTRICULAR HYPERTROPHY

Blood Pressure Profile → ← Genetic controls on myocyte response

RAAS tone →

← Exercise response

Adrenergic tone →

← Age-related change

Obesity →

Cellular trophic hormones → ← Concomitant IHD or valve disease

Outcomes

↑Myocardial wall stress
↓Endomyocardial perfusion
↑Likelihood of ischaemia infarction
↑Impairment of response to exercise
↑Risk of post infarction failure

Figure 9
LVH and hypertension: a bad combination and a role for the RAAS.

Although most, but not all, therapies that reduce blood pressure will cause reduction in LVH. There is more convincing evidence with the ACE inhibitors than with any other class of drug that they can induce regression. Other agents usually show a less marked effect on left ventricular mass, whereas the diuretics and calcium channel blocking drugs show an *inconsistent* pattern of effect in controlled studies.

Although there are as yet no independent prospective data to support reductions in morbidity and mortality from cardiovascular events by inducing regression of LVH, much preliminary data support this logical conclusion. In addition, although cardiac failure is a quite separate condition from hypertension, many trials of ACE inhibition in cardiac failure and following myocardial infarction show a reduction in the recurrence rate of myocardial infarction. Thus clinical evidence supports the efficacy of ACE inhibition from the general cardiac standpoint.

The reasons for this effect remain obscure but the possibilities are wide and do not exclude a primary effect. Indeed, during assessment of the adverse effect of nisoldipine on cardiac events within the ABCD (Appropriate Blood Pressure Control in Diabetes) study of hypertension in diabetic patients, there was a nine-fold better outcome (relative risk reduction) in the ACE inhibitor (enalapril) treated group.

In summary ACE inhibitors should probably be considered as at least one preferred therapy in patients with LVH evident during their assessment.

Positives and negatives in the use of ACE inhibitor therapy: how much of a balance is needed?

Blood pressure reduction is one part of the risk equation, as discussed above. Many groups of drugs can lower blood pressure in isolation and equivalence is easy to demonstrate on this basis alone in many clinical trials. Whether this is achieved in routine use depends very much more on the patient treated, their relationship to the physician, and the therapy chosen. The selection of ACE inhibition as a preferred tool to lower blood pressure in a particular patient is dependent on a broader canvas of factors (Figure 10).

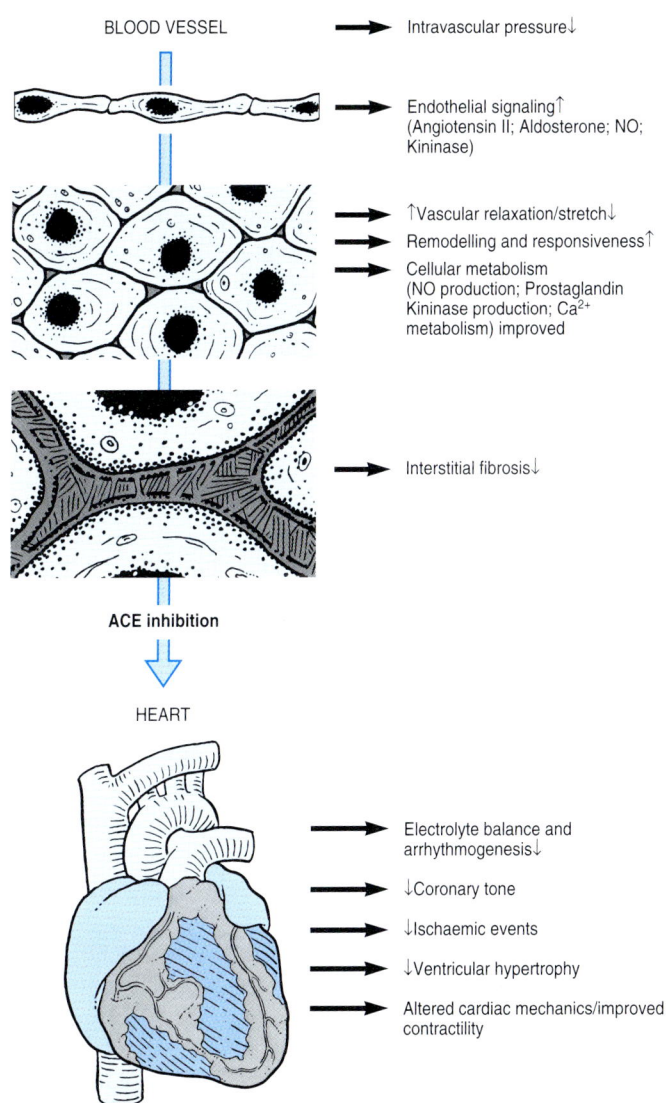

Figure 10
Mechanisms of vascular and cardiac protection by ACE inhibitor treatment.

Safety is critically important in hypertension, a condition affecting a third of people over 65 years throughout the world. ACE inhibition is undoubtedly safe in large numbers of clinical trials and in practical use. Caution is needed where pregnancy may be an issue and where a particularly marked response can be anticipated (see above). The RAAS is such a fundamental part of the cardiovascular system — in terms of not only blood pressure control but also the renal vascular and cardiac response to hypertension — that not using these agents becomes more the issue.

What are the negative aspects of ACE inhibitors? Initially one must look at the alternatives in drug treatment. Probably the most often discussed is simply a financial comparison, most often in relation to thiazide diuretic treatment. Thiazide diuretics have little potential for positive effects on the more general aspects of cardiovascular risk (eg, they tend to worsen lipid profile, promote electrolyte disturbance (Mg^{2+} and K^+) and aggravate hyperglycaemia in diabetics, albeit in a dose-related fashion and in only a minority of patients). If cost were not an issue, it is likely that few patients would receive diuretic therapy.

Beta blockade acts at another fundamental step in cardiovascular regulation but is often poorly tolerated from a symptomatic standpoint. In 1998 we must be concerned about the safety issue for the calcium channel blocking drugs — at least those in the dihydropyridine class (nifedipine, nisoldipine, etc). On a population exposure basis there seem to be some issues that we need to work through about their interaction in patients with ischaemic heart disease.

Mortality and cardiovascular event reduction in hypertension is the goal of therapy. There is no doubt that this level of evidence is missing in the treatment of mild to moderate hypertension with newer drug groups such as the ACE inhibitors.

This seems a remarkable omission after 20 years of use in hypertension. However, where mortality studies are relatively easy, eg, in heart failure due to the high rate of death, we are aware that such studies have overwhelmingly supported the use of ACE inhibitors as a fundamental treatment in this condition, dramatically distinct from experience with other vasodilators. Equally the use of ACE inhibition to attenuate the decline in renal function in diabetic and then non-diabetic hypertensive nephropathy has been well established, initially in experimental studies and later in patients. This again is in contrast to the effects of alternative blood pressure lowering strategies.

Given the indirect observational evidence already gathered, although it is still quite possible that blocking the RAAS is only equivalent to any other antihypertensive therapy, it would seem to require inverted logic (eg, the evidence regarding ACE inhibition is largely incorrect or biased and the evidence with other drugs is, by chance, less convincing and more accurate) to justify ignoring it. The mortality trials in hypertension should finish this debate early in the new millennium, but it is hard to believe that ACE inhibition would be seen as an inferior option. The question is only to what extent their increased benefit will be reflected in reduced mortality and events.

Which ACE inhibitor to use: is there an ideal candidate?

Diurnal blood pressure profile and cardiovascular pathology

ACE inhibition is therefore a therapy at the heart of cardiovascular disease prevention. As such it must be designed to cope with the onset of those diseases in the early hours of the morning. The diurnal profile of blood pressure has been documented since early studies of the beat to beat alterations in arterial pressure with the advent of intra-arterial monitoring of blood pressure. This is even easier to demonstrate with recent ambulatory non-invasive monitors. The early-morning rise in blood pressure is associated with sympatho-adrenal activation and probably occurs before waking or activity.

This surge in hormones and vascular stress is now felt to be important in vascular events and has to be counteracted (Figure 11). The hormones involved are likely, on the basis of recent evidence, to be the catecholamines, angiotensin and possibly aldosterone. ACE inhibitor therapy, though affecting all these systems to a differing degree, has to be effective at the end of its dose interval.

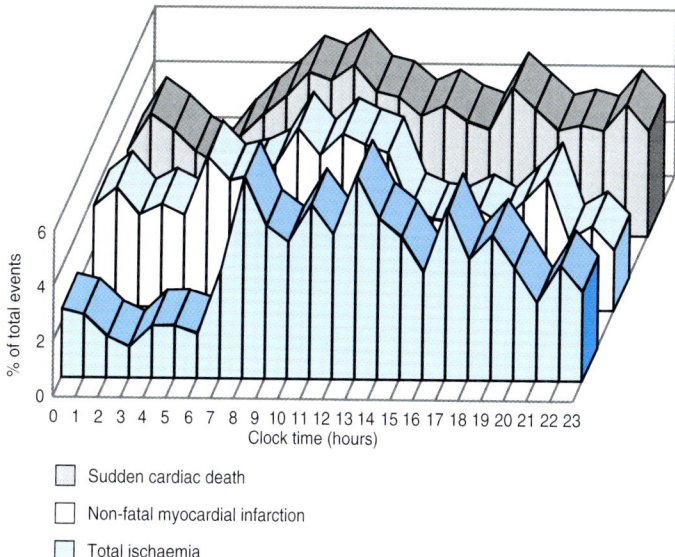

Figure 11
Circadian variation in transient myocardial ischaemia, non-fatal myocardial infarction and sudden cardiac death. (Reproduced with kind permission from D Mulcahy et al. Lancet 1988, ii: 755–9).

24 hours a day activity and the trough:peak ratio

The importance of 24 hour coverage and once-daily dosing outlined above has led to new assessments of blood pressure response. These are designed to describe the fluctuation of effect across the whole dose interval. This has been necessary to account for the use of excessive dose (and, with that, peak effect) in order to produce an acceptable trough effect compatible with once-daily administration. A smooth onset of drug response at any given dose, sustained throughout the dose interval, is preferred. This avoids overestimation of the trough effect in a drug that will have a low trough:peak ratio.

The trough:peak ratio is affected by a large range of variables dependent on the study design used, the patients treated and the calculation of the results, corrected for a placebo response to baseline blood pressure. Dose is a key factor in practice. Clearly, at the low doses used for initiation of therapy the trough:peak ratio can be low or even zero where drug effect is nil at the trough prior to the next dosing. The classical example of this within the ACE inhibitor group is captopril, for which the duration of effect is very short and the trough:peak ratio is accordingly zero.

In summary a profile of effect is sought where the trough:peak effect is ≥ 75% at the median dose used in clinical practice (Figure 12). This can be at the expense of peak blood pressure lowering effect, since it seems that it is the overall blood pres-

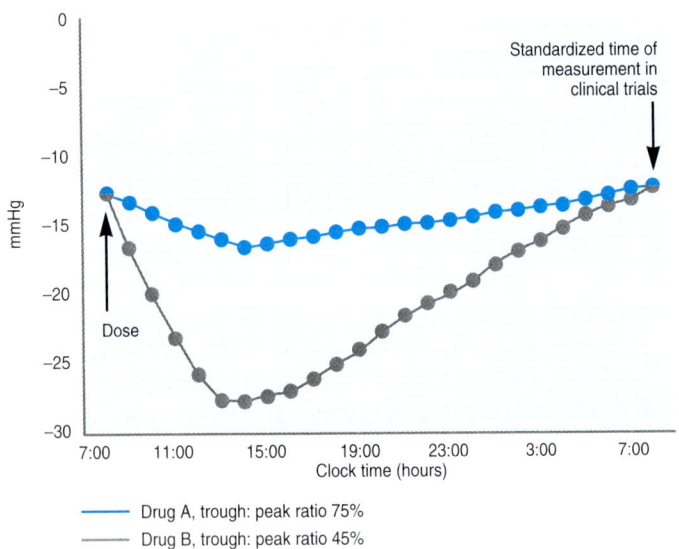

Figure 12
Trough to peak blood pressure response. (Reproduced from Meredith PA, Trough:Peak Ratio – What you need to know. In: Elliot HL ed, Current Issues in Cardiovascular Therapy, 19–33. Martin Dunitz: London, 1997.)

sure lowering over 24 hours that attenuates hypertensive organ damage. The most important aspect of this is in the reduction of LVH.

ACE inhibitor	% max BP effect remaining 24 hrs post-dose	Dose used
Perindopril	87–100	4 mg or 8 mg once daily
Enalapril	55–70	10 mg or 20 mg once daily
Captopril	38	50 mg twice daily

Table 2
24 hour action profile of ACE inhibitors. (From Morgan T et al. Clin Exp Pharmacol Physiol *1992;* **19 (suppl 19)***: 61–5; Morgan T et al.* Hypertension *1993;* **21***: 568.)*

Exercise control of blood pressure

Blood pressure varies throughout the day, not only in response to time and hormone signals as noted above but also with activity. This is most easily recognized in the form of the change in blood pressure between lying and standing or, more significantly, between rest and exercise. Although the benefits of control of resting blood pressure are established, it is known that some patients with end organ damage progress despite such control. Most hypertensive patients remain normally active and correctly are encouraged to exercise. Abnormal vascular and cardiac responses to activity (submaximal exercise) may persist despite apparently effective control of blood pressure at rest and may be one reason for the failure to reverse end organ damage despite effective control at rest.

The established effects of ACE inhibition on ventricular hypertrophy, vascular resistance and coronary blood flow are associated with moderate increases in cardiac output on long-term therapy. These effects are mediated by suppression of both angiotensin II and aldosterone, which is effective at rest and during exercise. Such therapy is likely to be effective in exercise suppression of excessive blood pressure response in active patients with hypertension.

Hormone suppression, blood pressure control and cardiac events

There is a link between the suppression of angiotensin (and probably aldosterone) and reductions in cardiac events in ischaemic heart disease and in heart failure. Similar relationships appear to exist in the hypertensive population when the setting of the RAAS is characterized in individual patients. The more active the RAAS, as defined very simply using renin activity and sodium excretion in the urine, the more cardiovascular events that occur in prospective follow-up (Figure 13).

Logically this suggests that suppression of these systems is important and also that the better the level of suppression the better the outcome should be. Equally, failure to suppress the effector hormones of the RAAS would result in a less successful outcome. This makes the link between events, the RAAS and the control of blood pressure using an agent blocking the RAAS. This role is effectively filled by a long-acting and potent ACE inhibitor drug.

The link between symptoms and compliance

Notwithstanding the above section on rare instances when symptoms are attributable to hypertension, the main factors affecting treated patients are the symptoms attributable to side-

Figure 13

Incidence of myocardial infarction, as adjusted for age, sex and race, according to renin profile and smoking status, cholesterol level or fasting blood glucose level. (Reproduced with permission from Alderman MH, Madhavan S et al. N Eng J Med 1991; **324**: 1098–104. Copyright ©1991 Massachusetts Medical Society. All rights reserved.)

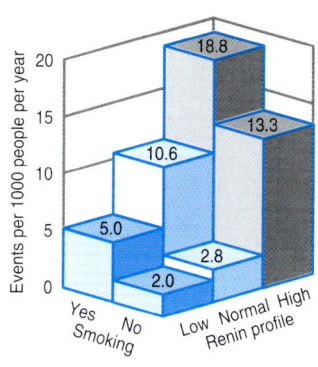

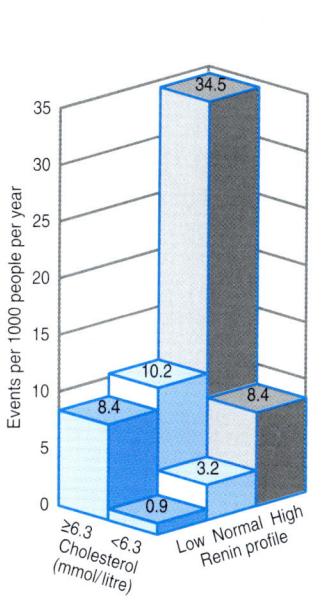

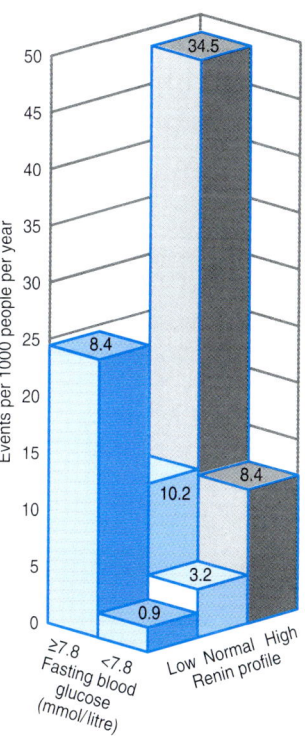

effects of drug therapy. This has become increasingly evident as a major issue in hypertension management. The main symptoms associated with each of the main classes of drug are well known (Table 3).

Within the ACE inhibitor group the side-effect profile is qualitatively similar but quantitatively different. The higher doses of captopril required for antihypertensive effect generally create more skin rashes and taste disturbance than do other agents. The typical dry cough of ACE inhibitor therapy seems to vary from < 1% to 10% of patients treated in different studies and does *not* seem to be dose related. Changing the agent may reduce or eliminate the cough in some studies; although dose reduction is advocated by some, the effects appear to be inconsistent.

The ACE inhibitor cough: all you ever needed to know

Why?
- Cough is a non-specific symptom and, though many people receiving ACE inhibition for hypertension will cough, only a small proportion will cough because of ACE inhibition

How?
- ACE inhibitors potentiate kinins in the body. These increase the sensitivity of pharyngeal and airway sensory receptors and create a dry non-productive cough in sensitive individuals

When/how often?
- Calculations vary dependent on the exclusion of non-specific (ie, non-drug-related) cough
- Cough will usually start at the onset of treatment or during dose titration
- Cough starting after many years of therapy should not be suspected to be drug related

What to do?
- The nature and possible origin of the symptom can be explained to the patient and they can decide if the side-effect merits any change in treatment
- Dose reduction can be tried dependent on blood pressure control

- Changing to an alternative ACE inhibitor can be effective although the reasons for this are unclear
- If blood pressure control is good on ACE inhibition and cough still proves problematic then an alternative blocker of the RAAS such as an angiotensin receptor antagonist can be tried. If cough persists despite this change then it is unlikely to be related to treatment

An ideal ACE inhibitor

1. Effective suppression

The properties of an ideal drug are easily summarized. Inherently, the key effector components of the system, angiotensin II and aldosterone, must be fully suppressed (i.e. absent). The drug should be able to provide suppression of the renin angiotensin system at rest and also during erect posture, and possibly also during activity or exercise.

2. Flexible dose-response

A dose-dependent range of suppression can be useful but does not need to have many incremental steps. The effectiveness of the drug should be demonstrable over 24 hours on the basis of the drug's pharmacokinetic profile rather than increments of dose.

3. Clinical impact

The blood pressure and ancillary tissue-dependent effects of the drug should be able to reverse key elements of the pathology of hypertension. Critical among these is the reversal of left ventricular hypertrophy and the attenuation of hypertensive neurotherapy. Combination effectiveness should be synergistic rather than additive.

4. Practical usage

The side-effect profile of the drug should be compatible with prolonged once daily use in an essentially asymptomatic population.

Class of drug	Symptoms	Psychomotor/ psychosensory functioning
Diuretics	Impotence, decreased libido, lethargy, constipation, nausea, dizziness	
β-adrenoceptor-blocking drugs	Dyspnoea, lethargy, dizziness, vivid dreams, cold extremities, blurred vision, blocked nose	Prolongation of complex reaction time, reduction of verbal memory, psychosensory impairment, depression
Centrally acting drugs		
Methyldopa	Diarrhoea, tiredness, weakness of limbs, weight gain, dry mouth, vivid dreams	Reduction of verbal memory
Clonidine	Tiredness, sleep disturbance, dry mouth, constipation	
Reserpine	Tiredness	Depression
ACE inhibitors	Skin rash, loss of taste, taste disturbance, cough	
Calcium antagonists	Constipation, dizziness, headache, nausea, flushing, ankle swelling	

Table 3
Frequently reported side-effects of antihypertensive drugs in large-scale placebo controlled trials. (Adapted with permission from Walker SW, Rossor RW (eds): Quality of life: Assessment and Application. *(Lancaster: MTP Press, 1988) pp 253–65.)*

Perindopril, a long-acting once-daily ACE inhibitor

Origins and chemistry

There are now many ACE inhibitor drugs licensed for clinical use. All are effective when taken regularly. The key features that differentiate them from each other are in their profile of activity and duration of effect. Ultimately these are defined by their physical chemistry and pharmacokinetic properties. Perindopril is one ACE inhibitor which has a set of properties that differentiate it from the other agents in its class. This has to do with its potency, duration of effect and ease of titration, which as we have seen above are all critical for clinical effectiveness in the management of hypertension.

Perindopril is a non-thiol, pro-drug ACE inhibitor which is activated by gastrointestinal esterases to the active drug perindoprilat. This agent has a long duration of action in blocking ACE/kininase II in plasma and body tissues (Figure 14). The relationship between the pro-drug ester and the active metabolite is an interesting one. This is a feature of most of the long-acting ACE inhibitors, such as enalapril, quinapril and trandolapril.

The interconversion from parent ester to active diacid ACE inhibitor is particularly important for perindopril, since the parent

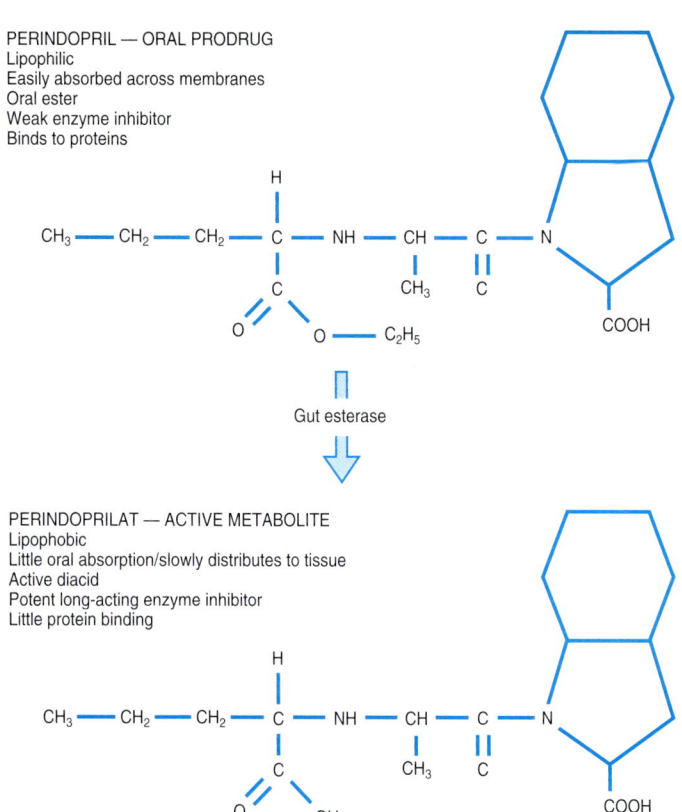

PERINDOPRIL — ORAL PRODRUG
Lipophilic
Easily absorbed across membranes
Oral ester
Weak enzyme inhibitor
Binds to proteins

CH_3 — CH_2 — CH_2 — C — NH — CH — C — N

Gut esterase

PERINDOPRILAT — ACTIVE METABOLITE
Lipophobic
Little oral absorption/slowly distributes to tissue
Active diacid
Potent long-acting enzyme inhibitor
Little protein binding

Figure 14
Perindopril and perindoprilat interconversion and efficacy.

drug perindopril has a higher affinity, yet weaker blocking capacity for ACE than perindoprilat. This has an important role in controlling the onset of enzyme inhibition in the blood but, more importantly, in the tissues and differs from the behaviour of even closely related drugs such as enalapril and their relationship to their active moieties such as enalaprilat. The result is that blood pressure lowering on initial dosing is gentle.

Profile of activity: hormone suppression, haemodynamic effect and dose response

The physical chemistry of perindopril and its relationship to perindoprilat are important aspects of its practical use (Figures 15 and 16). Because of this relationship, titration of dose is achieved in easy stages. An initial oral dose of 2 mg can be increased progressively by doubling to 4 mg and then 8 mg daily as a once-daily dose. At higher doses the increase in response rates is not substantial. Overall the response rate to monotherapy is good. Once-daily effectiveness can be demonstrated using ambulatory blood pressure monitoring.

Figure 15
Displacement binding of ACE inhibition by perindoprilat by the parent drug perindopril compared with little effect of enalapril on enalaprilat.

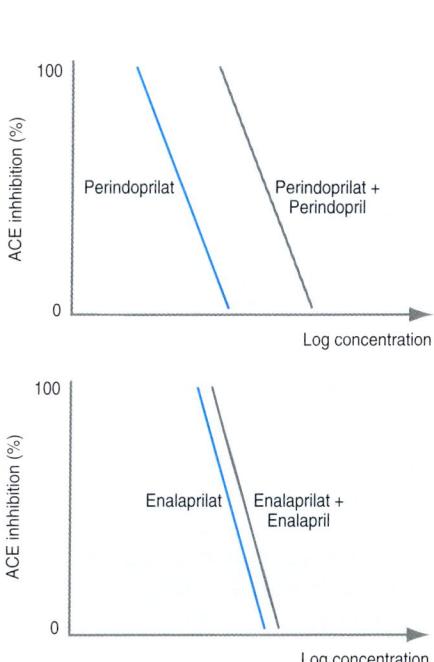

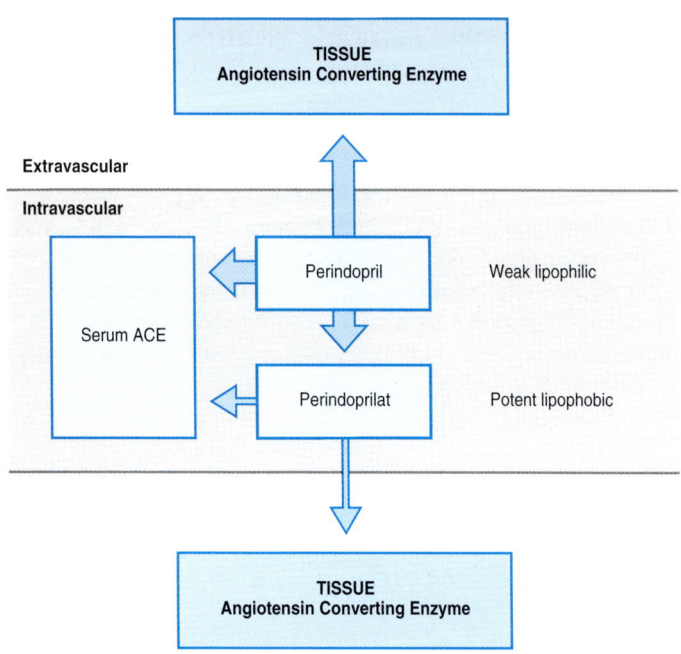

Figure 16
Perindopril–perindoprilat and tissue ACE: perindopril alternates the potent ACE inhibition of perindoprilat in serum and tissue. (First published MacFadyen RJ. Treating Heart Failure *1992;* **1***: 5–8.)*

The logical combination therapies with ACE inhibition are well known. They include the addition of a thiazide diuretic or, if a synergistic fall in blood pressure is required, a calcium channel blocking drug. Whatever the recent concerns over the use of the latter agents in hypertensive patients with ischaemic heart disease (IHD), they will remain very important adjuncts in the combination therapy of hypertension. While causing some suppression of renin, beta blockers tend not to provide such a useful additional fall in blood pressure.

Hormone suppression, as distinct from blood pressure reduction with ACE inhibitors, seems to be important in reducing cardiovascular event rates, certainly in heart failure and probably in hypertension. In this regard perindopril has an acceptable profile sustaining suppression of angiotensin II and aldosterone on chronic dosing. Reactivation of hormone production has been suggested to be a problem on chronic dosing with some ACE inhibitors but may relate to a variety of factors such as potency and duration of effect or, more likely, intermittent treatment compliance.

Safety profile and practical use

As with most major drugs post-marketing follow-up is essential to ensure the safety of an agent in day-to-day practical use. The safety profile of perindopril has been extensively documented in several large prospective studies in hypertensive patients.

Although the onset of response with perindopril is controlled, it should not be forgotten that this is a powerful vasodilator which blocks the RAAS to a high degree. Accordingly in situations where there is severe activation of the RAAS (through disease or excessive drug treatment with loop diuretics) or compensatory mechanisms (such as damage to the autonomic nervous system) are disrupted, then caution has to be exercised to balance the dose and response.

Perindopril is at the centre of a number of important comparative studies on large numbers of patients.

Cardiac events:
current and future cardiovascular trials

The use of ACE inhibition is established in patients with previous cardiovascular events. In particular we know that hypertensive patients with established ischaemic heart disease, cardiac failure or previous myocardial infarction, or significant end organ damage such as renal impairment should probably

Clinical trial	Patient group	Number of patients
ASCOT	Hypertension	9 000
EUROPA	Coronary artery disease	11 000
PROGRESS	Stroke	6 000

Table 4
Some current multicentre cardiovascular trials.

receive this form of treatment in preference to other antihypertensive drugs. This is based on previous multiple clinical trial experience with many agents in the class.

Given all this information, which points still need to be addressed to guide the clinician in the use of these agents in hypertension? Important clinical trials on vascular endpoints are still being done in order to extend the range of clinical indications (Table 4).

The ultimate trials of mortality and morbidity in response to the various drug treatment options in 'uncomplicated' hypertension are underway. These results remain very important to confirm that ACE inhibition should be more widely employed in the management of hypertension. So much can be done to help the hypertensive patient and we need to make good decisions on our patients' behalf. Selecting ACE inhibition as part of the treatment of one's patient is not only a good decision; it is one which is very easy to make with confidence of safety and efficacy.

Treatment groups	Outcome	Results due
Thiazide ± Atenolol vs Amlodipine ± Perindopril	Cardiac deaths	2004
Placebo vs ACE inhibitor	Cardiac deaths	2002
Placebo vs ACE inhibitor ± Diuretic	Recurrent stroke	2001

Further reading

Individual references are not provided here but a selection of easily obtainable contributions is listed. Any readers particularly wishing original references on specific points are more than welcome to contact the author directly on a personal basis. E-mail communication is preferred.

Contact:
Tel: (44) 1463 704000 ext. 5943
Fax: (44) 1463 711322
E-mail: r.macfadyen@abdn.ac.uk
      rjmacf@aol.com

Detailed major textbook
Brenner BM, Laragh JH. 1996 *Hypertension*. Baltimore: Williams & Wilkins, 1996.

Mid-range textbook
Kaplan NM. *Clinical Hypertension* 6th edn. Baltimore: Williams & Wilkins, 1994.

All-round practical clinical textbook
Beevers DG, MacGregor GA. *Hypertension in Practice*. London: Martin Dunitz, 1987.

Epidemiology of hypertension
Clausen J, Jensen G. Blood pressure and mortality: an epidemiological survey with ten years follow up. *J Hum Hypertension* 1992; **6**: 53–9.

White-coat response and ambulatory blood pressure measurement

Sheps SG, Pickering TG, White WB et al. Ambulatory blood pressure monitoring. *J Am Coll Cardiol* 1994; **23**: 1511–13.

Prasad N, MacFadyen RJ, MacDonald TM. Ambulatory blood pressure monitoring in hypertension. *Quart J Med* 1996; **89**:95–102.

Discontinuation of antihypertensive drug therapy

Jones JK, Gorkin L, Lian JF et al. Discontinuation of and changes in treatment after start of new courses of antihypertensive drugs: a study of a United Kingdom population. *Br Med J* 1995; **311**: 293–5.

Vaur L, Bobrie G, Dutrey-Dupagne C et al. Short term effects of withdrawing angiotensin converting enzyme inhibitor therapy on home self measured blood pressure in hypertensive patients. *Am J Hypertension* 1998; **11**: 165–73.

Blood pressure and job strain

Pickering T. The effects of occupational stress on blood pressure in men and women. *Acta Physiol Scand* 1997; **161**: 125–8.

Hypertension and cardiovascular risk

Wood D, de Backer G, Faergeman O et al. Prevention of coronary heart disease in clinical practice. Recommendations of the Second Joint Task Force of European and other Societies on Coronary Prevention. *Eur Heart J* 1998; **19**: 1434–1503.

Neaton JD, Wentworth D. Serum cholesterol, blood pressure, cigarette smoking and death from coronary heart disease. Overall findings and differences by age for 316,099 white men. *Arch Int Med* 1992; **152**: 56–64.

Zanchetti A, Sleight P, Birkenhager WH. Evaluation of end organ damage in hypertension. *J Hypertens* 1993; **11**: 875–82.

The renin angiotensin aldosterone system and hypertension
Fornage M, Amos CI, Kardia S et al. Variation in the region of the angiotensin converting enzyme gene influences inter individual differences in blood pressure levels in young white males. *Circulation* 1998; **97**: 1773–9.

Exercise limitation in hypertension
Lim PO, MacFadyen RJ, Shiels P et al. Exercise responses in white coat hypertension. *Lancet* 1996; **384**: 1445.

Lim PO, MacFadyen RJ, Clarkson PBM et al. Impaired exercise tolerance in hypertensive patients. *Ann Int Med* 1996; **124**: 41–55.

Antihypertensive therapy
Reid CM, Dart AM, Dewar EM. Interactions between the effects of exercise and weight loss on risk factors, cardiovascular haemodynamics and left ventricular structure in overweight subjects. *J Hypertens* 1994; **12**: 291–301.

Brunner HR, Menard J, Waeber B et al. Treating the individual hypertensive patient: considerations on dose, sequential monotherapy and drug combinations. *J Hypertens* 1990; **8**: 3–11.

Bulpitt CJ, Palmer AJ, Fletcher AJ et al. Optimal blood pressure control in treated hypertensive patients. *Circulation* 1994; **90**: 225–33.

Predicting antihypertensive response to ACE inhibition
Donnelly RJ, Meredith PA, Elliott HL. The description and prediction of antihypertensive drug response: an individualized approach. *Br J Clin Pharmacol* 1991; **31**: 627–34.

ACE inhibition and antihypertensive drugs in left ventricular mass regression
Fagard RH, Staessen JA, Thijs L. Relationships between changes in left ventricular mass and in clinic and ambulatory blood pressure in response to antihypertensive therapy. *J Hypertension* 1997; **15**: 1493–502.

Roman MJ, Alderman MH, Pickering TG et al. Differential effects of angiotensin converting enzyme inhibition and diuretic therapy on reductions in ambulatory blood pressure, left ventricular mass, and vascular hypertrophy. *Am J Hypertension* 1998; **11**: 387–96.

ACE inhibitor cough
Zee RY, Rao VS, Paster RZ et al. Three candidate genes and ACE inhibitor related cough — pharmacogenetic analysis. *Hypertens* 1998; **31**: 925–8.

ACE inhibition in renal disease
Navis G, de Zeeuw D, de Jong PE. ACE inhbitors: panacea for progressive renal disease. *Lancet* 1997; **349**: 1852–3.

Derkxx FHM, van Jaarsveld BC, Krijnen P et al. Renal artery stenosis towards the year 2000. *J Hypertens* 1996; **14 (suppl 5)**: S167–S172.

Perindopril in hypertension
MacFadyen RJ, Lees KR, Reid JL. Perindopril a review of its pharmacokinetics and clinical pharmacology. *Drugs 1990;* **39 (suppl 1)**: 49–63.

Morgan T, Anderson A, Jones E. The effect on 24 hour blood pressure control of an angiotensin converting enzyme inhibitor (perindopril) administered in the morning or at night. *J Hypertens* 1997; **15(2)**: 205–11.

Current and future cardiovascular trials
Neal B, MacMahon S. PROGRESS (perindopril protection against recurrent stroke study): rationale and design. *J Hypertens* 1995; **13**: 1869–73.

Index

Page numbers in *italics* refer to the illustrations.